I0471958

Highlights of GAO-12-447, a report to the Subcommittee on Defense, Committee on Appropriations, United States Senate

May 2012

TACTICAL AIRCRAFT

F-22A Modernization Program Faces Cost, Technical, and Sustainment Risks

Why GAO Did This Study

The Air Force currently plans to spend $11.7 billion to modernize and improve reliability of the F-22A, its fifth generation air superiority fighter. Originally designed to counter air threats posed by the former Soviet Union, the post-Cold War era spurred efforts to add new missions and capabilities to the F-22A, including improved air-to-air and robust air-to-ground attack capabilities. In 2003, the Air Force established the F-22A modernization program to develop and insert new capabilities in four increments.

GAO was asked to evaluate (1) cost and schedule outcomes and (2) testing results and risks going forward in the F-22A modernization program and related efforts. To do this, GAO examined the program's budgets and schedule estimates over time and discussed any changes with program officials, and reviewed progress and results from developmental and operational testing, and plans to mitigate risks and resolve system deficiencies.

What GAO Recommends

GAO recommends that DOD evaluate capabilities to determine if future F-22A modernization efforts meeting DOD policy and statutory requirements should be established as separate major acquisition programs.

DOD concurred with our recommendation.

View GAO-12-447. For more information, contact Michael J. Sullivan, (202) 512-4841, sullivanm@gao.gov

What GAO Found

Total projected cost of the F-22A modernization program and related reliability and maintainability improvements more than doubled since the program started–from $5.4 billion to $11.7 billion–and the schedule for delivering full capabilities slipped 7 years, from 2010 to 2017. The content, scope, and phasing of planned capabilities also shifted over time with changes in requirements, priorities, and annual funding decisions. Visibility and oversight of the program's cost and schedule is hampered by a management structure that does not track and account for the full cost of specific capability increments. Substantial infrastructure costs for labs, testing, management, and other activities directly support modernization but are not charged to its projects. The Air Force plans to manage its fourth modernization increment as a separate major acquisition program, as defined in DOD policy and statutory requirements.

Comparison of Estimated F-22A Modernization Program and Related Costs (Nominal Dollars in Millions, not Inflation Adjusted)

Then-year (dollars in millions)

Reliability improvements and structural repairs	Direct charges to modernization increments
Activities in support of modernization	

Source: GAO analysis of U.S. Air Force data.

Note: The 2004 estimate reflects costs from 2003 to 2012. The 2012 estimate reflects costs from 2003 to 2023.

Testing of new capabilities to ensure operational effectiveness and suitability is ongoing. Results to date have been satisfactory but development and operational testing of the largest and most challenging sets of capabilities have not yet begun. Going forward, major challenges will be developing, integrating, and testing new hardware and software to counter emerging future threats. Other risks are associated with greater reliance on laboratory ground tests and relocating an F-22A lab needed to conduct software testing. While modernization is under way, the Air Force has undertaken parallel efforts to improve F-22A reliability and maintainability to ensure life-cycle sustainment of the fleet is affordable and to justify future modernization investments. But the fleet has not been able to meet a key reliability requirement, now changed, and operating and support costs are much greater than earlier estimated.

_____ United States Government Accountability Office

Contents

Abbreviations

AT&L	Acquisition, Technology and Logistics
DCMA	Defense Contract Management Agency
DOD	Department of Defense
DOT&E	Director, Operational Test and Evaluation
DTM	Directive-Type Memorandum
IFDL	Intra Flight Data Link
JDAM	Joint Direct Attack Munition
OSD	Office of the Secretary of Defense
O&S	Operating and Support
MADL	Multifunction Advanced Data Link
MDAP	Major Acquisition Defense Program
MTBM	Mean Time Between Maintenance
RAMMP	Reliability and Maintainability Maturation Program

United States Government Accountability Office
Washington, DC 20548

May 2, 2012

The Honorable Daniel K. Inouye
Chairman
The Honorable Thad Cochran
Ranking Member
Subcommittee on Defense
Committee on Appropriations
United States Senate

The Air Force currently plans to spend about $11.7 billion to modernize and improve the reliability of the F-22A, the Air Force's fifth generation air superiority fighter. About $9.7 billion will be spent on specific modernization increments and related support costs, and nearly $2 billion will be used to improve the reliability of the F-22A and make structural repairs. Originally designed to counter air threats posed by the former Soviet Union, the post-Cold War era spurred efforts to add new missions and capabilities to the F-22A, including improved air-to-air and robust air-to-ground attack and capabilities. In 2003, the Air Force established the F-22A modernization program to develop and insert new capabilities. The timing and scope of the modernization program has changed over time, costs have significantly increased, and fielding of some capabilities has been delayed. In this context, you asked us to evaluate cost, schedule, and performance outcomes and risks of the F-22A modernization program.[1]

To determine the extent to which the F-22A modernization program is meeting cost and schedule goals, we researched the history of the program, plans and expectations at the start, and tracked budgets and schedule estimates over time. We identified changes in plans and estimates and discussed these changes with Air Force and Department of Defense (DOD) officials. To determine performance outcomes and risks remaining, we reviewed progress and results from developmental and operational testing and plans to mitigate risks and resolve system

---

[1]A companion report, GAO, *Tactical Aircraft: Comparison of F-22A Modernization Program and Legacy Fighter Modernization Programs,GAO-12-524* (Washington, D.C.: Apr. 26, 2012) addresses how the timing and strategy of the F-22A modernization program compares to similar past efforts on legacy fighter programs, including the Air Force's F-15 and F-16 and the Navy's F/A-18.

deficiencies. This included reviewing and discussing annual test reports from DOD's Office of the Director, Operational Test & Evaluation (DOT&E), briefings to defense oversight and requirements offices, summaries of recent operational test results provided by Air Force test officials, and program risk information related to testing new capabilities.

We conducted this performance audit from June 2011 to April 2012 in accordance with generally accepted government auditing standards. Those standards require that we plan and perform the audit to obtain sufficient, appropriate evidence to provide a reasonable basis for our findings and conclusions based on our audit objectives. We believe that the evidence obtained provides a reasonable basis for our findings and conclusions based on our audit objectives.

Background

The Air Force's F-22A Raptor is the only operational fifth-generation tactical aircraft, incorporating a low observable (stealth) and highly maneuverable airframe, advanced integrated avionics, and a supercruise engine capable of sustained supersonic flight. The F-22A acquisition program began in 1991 with an intended development period of 12 years and a planned quantity of 648 aircraft. The system development and demonstration period eventually spanned 14 years, during which time threats, missions, and some requirements changed. In particular, the F-22A was originally designed to fly primarily air-to-air missions; however, since that time the Air Force has decided to add air-to-ground capabilities to the F-22A. Development costs substantially increased and total quantities were eventually decreased to 188 aircraft. When the final aircraft is delivered in May 2012, the F-22A acquisition program will be complete at a cost of $67.3 billion.

In 2003, the Air Force established a modernization program to develop and insert new and enhanced capabilities considered necessary to meet the threat. According to Air Force officials, modernization is defined as a process of upgrading and modifying aircraft with a focus on adding new capabilities. The modernization is now proceeding in four related increments, each with multiple projects:

- Increment 2,[2] the initial phase of modernization, addressed some requirements deferred from the acquisition program and added some new ground attack capability. It has been fielded.
- Increment 3.1 began fielding in November 2011 and adds enhanced radar and enhanced air-to-ground attack capabilities.
- Increment 3.2A is a software upgrade to increase the F-22A's electronic protection, combat identification, and Link-16 communications and data link capabilities.
- Increment 3.2B will increase the F-22A's electronic protection, geo-location, and Intra Flight Data Link (IFDL) capabilities, and adds AIM-9X and AIM-120D missiles.

In addition to these efforts, in 2006, the Air Force began a Reliability and Maintainability Maturation Program (RAMMP). Although the Air Force does not consider this part of the modernization program, it is integral to making the F-22A weapon system more available, reliable, and maintainable. Since the F-22A's initial fielding in 2006, maintenance issues have prevented it from achieving reliability and availability requirements, and fleet operating and support (O&S) costs are much higher than projected earlier in the program.

F-22A Modernization Costs Have Increased and Deliveries of New Capabilities to the Warfighter Have Been Delayed

Total projected cost of the F-22A modernization program has more than doubled since it started. While the program has completed and fielded some of its planned capabilities, the overall schedule to complete integration and testing of planned capabilities and deliver them to the warfighter has slipped by nearly 7 years. The content, scope, and phasing of planned capabilities also shifted over time with changes in requirements, priorities, and annual funding decisions. Visibility and oversight of the program's cost and schedule is hampered by a management structure that does not directly track and account for the full cost of specific capability increments. The Air Force plans to separately break out and manage the fourth increment as a major defense acquisition program, which should improve management and oversight.

---

[2] The Air Force numbering scheme considers increment 1 to be the baseline capabilities delivered by the F-22A acquisition program.

F-22A Modernization Costs Have Risen Sharply Since the Program Began

The Air Force is now expected to spend around $11.7 billion to modernize and improve the reliability of the F-22A, compared with the $5.4 billion projected soon after the start of development. Officials underestimated the scope of the total program and the time and money that would eventually be needed to develop and field new capabilities. Contributing factors to this cost growth include (1) changed and added requirements; (2) unexpected expenses for building a support infrastructure; and (3) unplanned efforts to improve aircraft reliability and maintainability. Program officials also said that instability in modernization funding contributed to some of the cost growth by stretching the time required to complete projects. Figure 1 shows increased cost estimates over time for the modernization program and other related costs.

Figure 1: Comparison of Estimated F-22A Modernization Program and Related Costs (Nominal Year Dollars in Millions, not Inflation Adjusted)

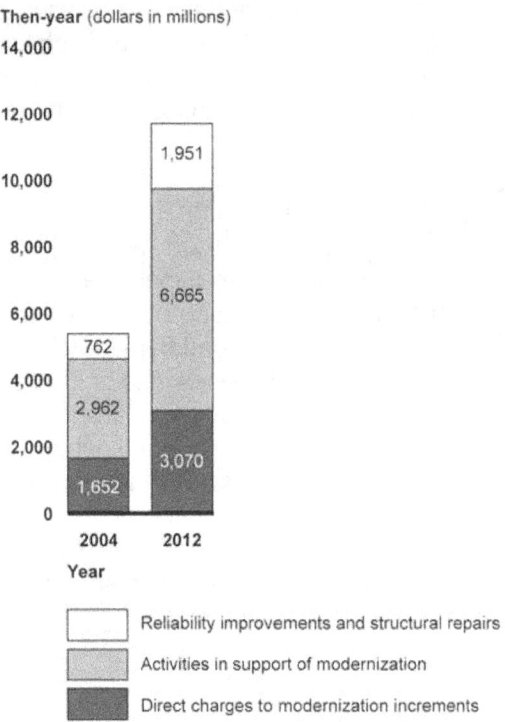

Source: GAO analysis of U.S. Air Force data.

Note: The 2004 estimate reflects projected costs from 2003 to 2012. The 2012 estimate reflects costs from 2003 to 2023.

Modernization increments include development and procurement costs directly tied to one of the four increments for acquiring upgraded capabilities. These include infrastructure costs for lab support, test operations, program management, retrofit to bring all aircraft to a common configuration, and other efforts integral to supporting modernization increments. Other improvement costs principally include the RAMMP reliability and maintainability projects and making structural repairs needed for the aircraft to achieve its required 8,000 hour service life. At this point, an estimated $5.5 billion of the $11.7 billon has been spent. A future investment of around $6.2 billion remains: $1.3 billion for Increment 3.2B, $3.6 billion for other modernization and support activities, and $1.3 billion for completing the RAMMP and structural repairs.

Completion of F-22A Modernization Projects Has Been Significantly Delayed as Content, Scope, and Phasing of Capabilities Changed over Time

When the F-22A modernization development program began, the Air Force expected to have all current planned capabilities integrated and fielding started by 2010. Now, the final increment is not expected to begin fielding until 2017, 7 years later than initially planned. Air Force officials stated that they underestimated the sheer magnitude of the modernization effort, both in the amount of time required to develop and integrate the capability, and costs to complete the modernization. According to program officials, contributing factors to delays include (1) additional requirements, (2) unexpected problems and delays during testing, and (3) research, development, testing, and evaluation funding fluctuations. Figure 2 compares the initial and latest schedules.

Figure 2: Initial and Latest F-22A Modernization Schedules (as of December 2011)

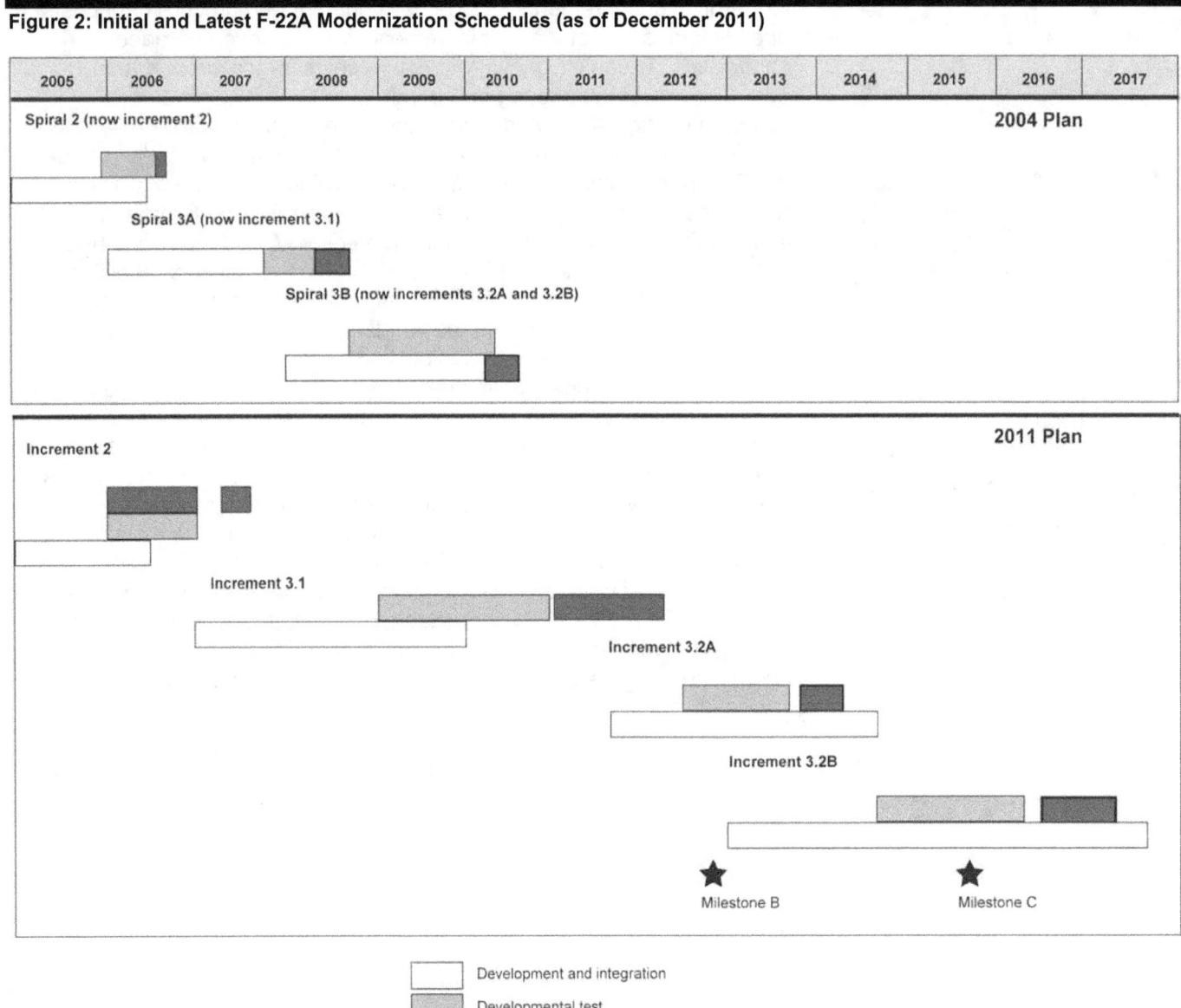

Source: GAO analysis of U.S. Air Force data.

Note: The Air Force has now replaced the term spiral development with increments which more accurately defines their approach to modernization. DOD Instruction 5000.02, Operation of the Defense Acquisition System (Dec. 8, 2008), provides for three decision points or phases. They include: milestone A (entry point for the technology development phase); milestone B (entry point for engineering and manufacturing development phase—which is comprised of two major efforts called integrated system design, and system capability and manufacturing process demonstration); and milestone C (entry point for the production and deployment phase).

According to Air Force officials, the program currently intends to upgrade 143 aircraft with the full complement of modernized capabilities by fiscal year 2020 and retain 36 aircraft with only Increment 2 capabilities to be used in training. Increment 3.1 is being fielded in fiscal years 2011 to 2016 and Increment 3.2A from fiscal years 2014 to 2016. Increment 3.2B, the last currently planned increment, is expected to field from fiscal years 2017 and 2020. Future capability enhancements are expected to follow the current modernization program, but have not been defined.

The content, scope, and phasing plan changed over time, contributing to cost and schedule problems. Figure 3 illustrates the changing nature of modernization projects, particularly in the later increments.

Figure 3: F-22A Modernization Planned Capabilities in Increments Have Changed over Time

Increment 2	2005	2008	2011
Candidate 38A	●	●	●
Chaff Track Delete	●	●	●
Full Chaff Fix	●	●	●
Intra-Flight Data Link Widenet	●	●	●
Launch Zone Improvement	●	●	●
Over-the-Horizon Data Transmit	●	○	○
Supersonic JDAM	●	●	●

Increment 3.1	2005	2008	2011
Air to Ground Radar	●	●	●
Airborne Network Transmit	●	○	○
Electronic Attack	●	●	●
Geolocate 1	○	●	●
Geolocate 2	●	○	○
JDAM Self Target	●	●	●
SDB Basic	●	●	●

Increments 3.2A and 3.2B	2005	2008	2011
AIM-9X[b]	○	●	●
AIM-120D[b]	●	●	●
Automatic Ground Collision Avoidance System	●	●	○
Autonomous Search	●	○	○
Combat ID[a]	○	●	●
Electronic Protection Update[a&b]	●	●	●
Enhanced Data Recording	●	○	○
Geolocate 1.5	○	●	○
Geolocate 2[b]	○	○	●
Geolocate 3 and 3A	●	○	○
Ground Moving Target Indication and Tracking	●	○	○
IFDL Upgrades[b]	○	○	●
Multi-function Advanced Data Link	○	●	○
Multi-function Radar	●	○	○
Link 16 Upgrades[a]	○	●	●
SDB Full	●	●	○
Selective Available and Anti-Spoofing Module	●	○	○

● Included in increment as of this year

○ Not included in increment as of this year

Source: GAO analysis of U.S. Air Force data.

Some capabilities, such as the Multifunction Advanced Data Link, have been eliminated because of changes in requirements and immature technology. Some, like the AIM-9X missile, have been added to the program to meet emerging threats. Some required capabilities have been reduced, such as the Geolocate project, which will now field a less-capable version than initially planned.

Air Force officials stated that potential new capabilities are analyzed and vetted by evaluating technical maturity and applying cost as independent variable principles[3] to determine which to include in the F-22A modernization program. As a result of this evaluation process, certain capabilities have been modified, deferred, added, or eliminated. Most changes affect the final two increments. For example, MADL, which was intended to provide communications interoperability with the F-35 Joint Strike Fighter, was removed from Increment 3.2B. MADL and other deferred efforts, such as the full Small Diameter Bomb capability, may eventually be delivered in future increments yet to be defined.

Visibility and Oversight of the Program's Cost and Schedule Is Hampered by a Management Structure and Funding Mechanism

Tracking and accounting for the full and accurate cost of each modernization increment, and individual projects within each increment, are limited by the way the modernization program is structured, funded, and executed. As depicted in figure 4, only 26 percent of total projected costs can be traced directly to the four modernization increments. About 57 percent of total costs go to fund activities that support all the modernization efforts and the overall F-22A program but are not charged to specific increments. These activities include test operations, the building and use of government labs, management activities, retrofit efforts to bring the fleet to a common configuration, and other infrastructure accounts. The remaining 17 percent funds the RAMMP program and structural repairs. While Air Force officials do not consider these efforts as part of the funded modernization program, we note that

[3]Cost as an independent variable principles call for the establishment of cost goals for operations, sustainment, and procurement, and for acquisition programs to make trade-offs in terms of cost, schedule, and performance.

these efforts are needed to improve fleet affordability and achieve desired aircraft life and thus integral to justify future modernization investments.

Figure 4: Air Force Current Estimate of F-22A Modernization and Other Improvement Costs

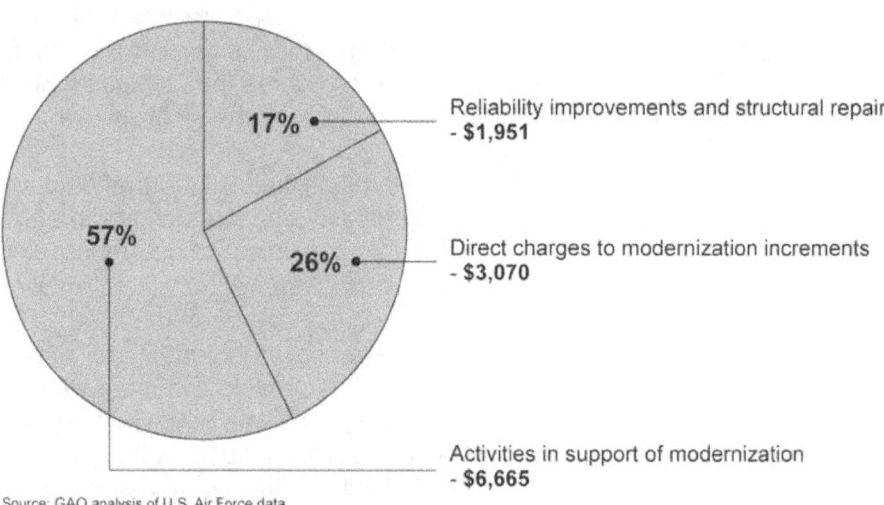

Dollars in millions

17% — Reliability improvements and structural repairs - $1,951

26% — Direct charges to modernization increments - $3,070

57%

Activities in support of modernization - $6,665

Source: GAO analysis of U.S. Air Force data.

Program accountability and oversight have been hampered by how the modernization program was established, managed, and funded. As we reported in March 2005, the Air Force embarked on the modernization program without a knowledge-based business case to support the multibillion dollar investment to significantly change the aircraft's capabilities and missions.[4] We stated that the modernization program should have been established as an entirely separate acquisition program with a new business case because of the magnitude of the proposed changes. A sound business case would have matched requirements with resources—proven technologies, sufficient engineering capabilities, time,

---

[4]GAO, *Tactical Aircraft: Air Force Still Needs Business Case to Support F/A-22 Quantities and Increased Capabilities*, GAO-05-304 (Washington, D.C.: Mar. 15, 2005).

and funding—when undertaking new product development.[5] However, information about the schedule and funding was not adequately known at the start of modernization. Rather than making the new business case to justify and manage the modernization program as a separate major defense acquisition, Air Force officials incorporated it within the existing F-22A acquisition program and comingled funds. Their rationale was their belief that breaking modernization efforts out as a separate program would have delayed the capability. As a result, development funding and infrastructure expenses were added to the existing acquisition program's baseline.

The modernization program proceeded without establishing its own set of acquisition milestones and has not been subject to the same level of scrutiny by senior defense leaders or the performance reporting required of major defense acquisition programs as provided for in DOD acquisition policy. At their discretion, DOD chose to execute it within the baseline F-22A program. [6]

In November 2004, defense leaders recognized that the size and importance of the modernization program warranted a higher level of scrutiny. The acting Under Secretary of Defense for Acquisition, Technology and Logistics directed the Air Force to hold separate milestone reviews for the future stages of the modernization program to be consistent with DOD acquisition policy. Under this Air Force direction, the current modernization projects would not require formal milestones,

---

[5]A business case is defined as demonstrated evidence that (1) the warfighter need exists and that it can best be met with the chosen concept and (2) the concept can be developed and produced within existing resources—including design knowledge, demonstrated technologies, adequate funding, and adequate time to deliver the product, GAO, *Defense Acquisitions: Managing Risk to Achieve Better Outcomes,* GAO-10-37T (Washington, D.C.: Jan. 20, 2010).

[6]Currently, a major defense acquisition program (MDAP) is a DOD acquisition program that is not a highly sensitive classified program and that is designated by the Under Secretary of Defense for AT&L as a MDAP or that is estimated to require an eventual total expenditure for research, development, test and evaluation, including all planned increments, of more than $365 million (based on fiscal year 2000 constant dollars) or an eventual total expenditure for procurement, including all planned increments, of more than $2.9 billion (based on fiscal year 2000 constant dollars). Directive-Type Memorandum (DTM) 09-027, *Implementation of the Weapon Systems Acquisition Reform Act of 2009* (Dec. 4, 2009, incorporating change 3, Dec. 9, 2011), Attachment 1 at § 13. See also, DOD Instruction 5000.02, *Operation of the Defense Acquisition System* (Dec. 8, 2008) and 10 U.S.C. § 2430.

but Office of the Secretary of Defense (OSD) oversight would be provided by periodic reviews. In 2007, OSD directed the Air Force to update the F-22A Acquisition Program Baseline to reflect the approved Increments 3.1 and 3.2; however, the Air Force believed that an acquisition strategy report rather than a baseline would provide better insight into funding and schedule details for the modernization increments. Not separating the modernization program from the F-22A program baseline was consistent with how the Air Force had handled modernization programs for prior aircraft. However, had the Air Force initiated the program under existing guidelines established by DOD Instruction 5000.02 for managing and implementing major acquisition programs, oversight of the program would have benefitted. Under these guidelines, programs are required to have an approved minimum set of Key Performance Parameters, included in the Capability Development Document; an approved Acquisition Strategy; Acquisition Program Baseline; an Analysis of Alternatives; and an Independent Cost Estimate, for a Milestone B decision that would allow them to proceed into the Engineering and Manufacturing Development phase.

OSD recently reiterated its requirement for the F-22A to be consistent with DOD policy, and in December 2011, OSD directed the Air Force to establish increment 3.2B as a separate major defense acquisition program. According to the Air Force, this increment is expected to cost around $1.5 billion. Given the significant slips in schedule experienced by increments 3.1 and 3.2A, the decision to separately oversee increment 3.2B is a late but positive change. Increment 3.2B will be reported as its own major program with system development starting in fiscal year 2013. This should improve management, cost visibility, and program oversight. Air Force officials told us that they expect to manage and report all future F-22A modernization programs as separate acquisitions, starting with Increment 3.2B.

Performance Outcomes Have Been Judged Satisfactory, but Testing and Improving Reliability and Affordability of the Fleet Will be Challenging

Testing how well new capabilities perform is ongoing; results to date have been satisfactory but development and operational testing of the largest and most challenging sets of capabilities have not yet begun. Going forward, major challenges will be developing, integrating, and testing new hardware and software to counter emerging future threats. Other risks are associated with availability of unique test assets, greater reliance on laboratory ground tests, and relocation of a key F-22A lab that is needed to help support testing of software for the new capabilities. Parallel efforts to improve F-22A reliability and maintainability are critical to ensure life-cycle sustainment of the fleet is affordable and to justify future modernization investments.

Operational Testing Results on New Capabilities Have Been Mostly Positive to Date, but More Challenging Efforts Are Still Ahead

New F-22A capabilities delivered by the modernization program will be demonstrated through follow-on operational testing and evaluation to assess the upgraded F-22A's effectiveness and suitability.[7] Testing on the first two increments successfully demonstrated new air-to-ground capabilities. Testing of the third and fourth increments has not begun and several technical risks remain for these new capabilities. Successful mitigation of these risks is critical to keeping F-22A's planned upgrades on schedule and within planned costs. Table 1 shows the current status of F-22A modernization operational testing for each increment.

---

[7]Operational effectiveness is the overall degree of mission accomplishment of a system when used by representative personnel in the environment planned or expected for operational employment and operational suitability is the degree to which a system can be satisfactorily placed in field use, with consideration given to reliability, availability, and maintainability.

Table 1: Current Status of F-22A Modernization Operational Testing for Each Increment, as of December 2011

Increment	Capabilities tested	Key issues during operational testing	Operational testing completed
Increment 2	Fixes to baseline aircraft and expanded air-to-ground capabilities against fixed targets with 1,000 pound Joint Direct Attack Munition.	Inspection and repair of low observable components required significant effort, and accounted for half of all maintenance hours.	August 2007
Increment 3.1	Additional air-to-ground capabilities allowing the F-22 to find and fix targets by itself, without the need for external platforms to provide coordinates.	Fleet stand-down due to concerns about the oxygen generation system; unavailability of a test range for flight testing and technical delays related to ground support equipment. Operational requirement for reliability was changed during testing.	November 2011
Increment 3.2A	Updates to electronic protection, combat identification and targeting capabilities.	Operational testing is expected to begin in late 2013.	Early 2014 (estimated)
Increment 3.2B	Improved strike capabilities with AIM-9X and AIM-120D missiles, and more advanced geolocation and electronic protection capabilities.	Operational testing is expected to begin in late 2016.	Mid-2017 (estimated)

Source: GAO analysis of DOD data.

Increment 2

Follow-on operational testing and evaluation for F-22A fighters incorporating Increment 2 capabilities, including assessments of expanded air-to-ground capability and improvements in system suitability, were successfully completed in August 2007. The F-22A's configured with Increment 2 capabilities were found to be operationally effective in suppressing and destroying fixed enemy air defenses, and also demonstrated successful fixes of deficiencies and weapons integration problems that had caused problems in previous testing. Flight testing demonstrated the ability to employ the Joint Direct Attack Munition (JDAM) at supersonic speeds in a high- threat anti-access environment where stealth capabilities are needed. Without this capability, baseline aircraft were only able to launch JDAMs at fixed targets in lower threat environments and at slower speeds while using target coordinates from ground spotters.

Increment 3.1

Increment 3.1 further enhances F-22A's air-to-ground capability by allowing the aircraft to find and locate ground targets with on-board systems, rather than relying on external personnel and platforms for targeting. Increment 3.1 completed follow-on operational testing in November 2011; a significant delay of 4 years from the original plan due to shortcomings identified with the baseline and upgraded aircraft. In 2009 and 2010, the Director of Operational Test and Evaluation (DOT&E)

reported significant stealth-related maintenance issues that lowered operational availability and mission capability rates. F-22A program officials identified technical issues in upgrading radar, navigation, and software that needed to be addressed to meet operational testing requirements.

The Air Force began Increment 3.1 operational testing in January 2011, but soon encountered flight delays that persisted from March to September 2011. The entire F-22A fleet was ordered to stand-down due to potential problems with the aircraft's oxygen generation system. Unavailability of the test range and technical problems with ground support equipment also contributed to the lengthy flight delay. The Air Force completed flight testing for Increment 3.1 in November 2011 and expects to release the operational test report in late March 2012. In its 2011 annual report, DOT&E did not identify any significant remaining issues since flights had resumed. DOT&E also approved reducing trials from 16 to 8 and decreasing simulator test trials from 96 to 64. According to program officials, hardware and software issues had been identified and fixed as testing progressed and test pilots provided very positive feedback on Increment 3.1's enhancements.

Increment 3.2A

Increment 3.2A development began in November 2011 after significant delays. This increment involves updating software to enhance electronic protection and combat identification capabilities, so that F-22A can handle new threats expected in the future. Developmental testing for this increment is expected to start in 2012 and be completed in late 2013. Operational testing and evaluation will follow and is planned to conclude in 2014.

Program officials assessed the Increment 3.2A schedule as having moderate risk. Test aircraft have been operating much longer than planned and were to be replaced by new production aircraft; however, this has not happened due to the substantial reduction in the size of the F-22A fleet. Other risks appear lower. For example, some software for electronic protection and combat identification capabilities has already been developed for the F-35 Joint Strike Fighter. Also, while the Link-16 upgrade will involve a significant amount of development work, program officials consider it to be moderate risk.

Increment 3.2B

Increment 3.2B is scheduled to begin Engineering and Manufacturing Development in December 2012 and the decision to enter into production is scheduled for January 2016. Key capability upgrades include integrating the AIM-9X and AIM-120D missiles on the F-22A and upgrading geolocation and electronic protection subsystems. Early requirements analysis determined that AIM-9X integration may be more difficult and take longer than expected and officials have already begun risk reduction efforts. Overall, software integration is considered to have the highest risk for Increment 3.2B projects, while hardware development is rated as a moderate risk. Program officials believe that the full range of capabilities added in the modernization program can be accommodated within the weight and space limitations of the F-22A aircraft, but this will be a critical consideration in any future modernization plans.

The Air Force is seeking ways to reduce the costs of Increments 3.2A and 3.2B by streamlining program activities. Officials want to make more use of developmental tests to also satisfy operational test requirements, allowing the program to identify errors for correction earlier and reducing overall costs by eliminating redundant tests. The program also intends to increase its use of F-22A ground laboratories to substitute for more expensive flight tests. The F-22A lab infrastructure is an extensive, distributed system of dedicated labs that integrate and certify flight software releases to the field and support F-22A modernization, production and sustainment activities. However, there are technical risks if lab tests do not fully replicate the performance of actual F-22A aircraft in intended environments. Officials are also expecting to save money by relocating the Raptor Avionics Integration Lab—a critical work site that stimulates sensors for targeting—from Marietta, Georgia, to Ogden Air Logistics Center, Utah by the summer of 2012. Program officials acknowledge there are some risks in this. For example, unique equipment could be damaged during the move and experienced lab staff could decide to leave the F-22A program rather than relocate.

RAMMP Program Addresses Reliability and Maintainability Deficiencies to Improve Affordability and Justify Future Investments

In addition to capability upgrades, the F-22A budget also funds efforts to address reliability and maintainability deficiencies that have increased support costs and have prevented the F-22A from meeting a key performance requirement. RAMMP is to develop and implement enhancements to increase aircraft availability, make maintenance faster and less costly, and reduce total life-cycle operating and support costs and cost per flying hour. While RAMMP is expected to reduce life-cycle costs over the long term, up-front investments to help realize future cost reductions have increased. The program had planned to spend about $258 million between 2005 and 2011, but actual investments through 2011 were about $528 million. The total RAMMP funding requirement through the year 2023 is now estimated at almost $1.3 billion. Air Force officials attributed part of RAMMP's increased costs to additional projects and increased labor hours to address corrosion.

Keeping the F-22A fleet affordable and meeting required performance measures is critical to sustaining fleet operations over the long term, and ensuring it is available in sufficient numbers for required missions. Projected operational and support costs are much higher than earlier estimates. For example, a 2007 independent estimate by the Air Force Cost Analysis Agency projected a $49,549 cost per flying hour in 2015 (by which time the F-22A was expected to reach full maturity), more than double the $23,282 cost per flight hour estimated in 2005.

Air Force officials gave various reasons for sustainment cost increases including (1) unrealized savings from the F-22A's performance-based logistics contract[8] (2) fixed costs that had to be spread over a smaller number of aircraft; and (3) higher than expected costs to refurbish or replace broken parts, including diminishing manufacturing sources. However, the one common contributing factor—and the most impactful— is the cost and complexity of maintaining stealth characteristics and restoring aircraft to the required stealth level after flight operations and maintenance. Our recent report found that the number of maintenance personnel required to maintain the F-22A's specialized stealth exterior has increased, posing a continuing support challenge for this aircraft.[9] This has important implications for the affordability and life-cycle cost estimates for the F-35 Joint Strike Fighter.

When it started in 2006, a major goal of RAMMP was to improve F-22A reliability to meet its key performance requirement by the time the fleet reached maturity at 100,000 total flight hours. This performance indicator, known as mean time between maintenance (MTBM), required aircraft in the F-22A fleet to fly an average of 3 hours between maintenance events, excluding routine servicing and inspections. This performance standard was a key performance requirement in the F-22A acquisition contract, but the fleet has never been able to meet that requirement. Currently, the

---

[8] The F-22A's prime contractor, Lockheed-Martin, provides life-cycle product support, including supply and maintenance under this arrangement.

[9] GAO, *Defense Management: DOD Needs Better Information and Guidance to More Effectively Manage and Reduce Operating and Support Costs of Major Weapon Systems*, GAO 10-717 (Washington, D.C.: July 20, 2010).

MTBM achieved by the operational test aircraft with improvements is 2.47 hours.

In April 2011, the Joint Requirements Oversight Council approved changing the main reliability metric from MTBM to another performance indicator, known as material availability. Officials believed the MTBM indicator was hard to define and measure, was unrealistic, and did not accurately reflect the fleet's readiness to perform missions. Material availability is defined as the percentage of the fleet available to perform assigned missions at any given time. This standard calls for the F-22A fleet to achieve increasing levels of availability between 2011 and 2015 toward the final goal of 70.6 percent. Last year, the F-22A fleet achieved a 55.5 percent materiel availability rate. Stealth-related maintenance, system component reliability problems, and lack of spare engines were factors contributing to the fleet not achieving the goal. However, program officials expect the F-22A fleet to achieve the final availability goal by 2015 after the full fielding of reliability improvements. The Air Force reported that operational test on aircraft integrated with the current reliability improvements have achieved 78 percent availability; they anticipate significant gains by the overall fleet once reliability improvements are installed on all F-22A aircraft.

Conclusions

Keeping the F-22A as the world's most advanced stealth fighter requires the Air Force to counter changing threats, as well as ensure the F-22A fleet is affordable, reliable, and sustainable. In response to changing threats, officials began a Modernization Program to add new missions and capabilities while fixing problems and deficiencies that were carried over from the original development program. However, the F-22A modernization program has not had the management rigor or oversight on par with the $11.7 billion investment it entails. The program was not well-defined when it began in 2003, has had fluid scope and cost, and has been challenging from an oversight perspective as it was blended into the baseline F-22A program rather than being managed separately. As early as 2004, OSD began discussing the need to manage future modernization increments as separate acquisition programs. While modernization has been underway, the Air Force has found it necessary to invest in improved reliability and availability of the F-22A through the RAMMP program. The original reliability requirement was not met and has since been changed to another indicator. Meanwhile, O&S costs have been significantly higher than planned, with maintenance of the aircraft's stealth levels being particularly demanding. The lessons learned on the maintenance of the stealthy F-22A may have implications for the F-35 Joint Strike Fighter.

Splitting out increment 3.2B as a separate major acquisition defense program indicates that OSD is reasserting its role in the F-22A program. This is beneficial for oversight in light of the significant decisions and investments yet to come for the program. Increment 3.2B requires around $1.3 billion, while completing the RAMMP program, ongoing modernization projects, and other improvements will require an estimated $4.9 billion—a total future investment of around $6.2 billion. The program is highly dependent on a single contractor, whose responsibilities encompass managing the development and production of the F-22A; development, production, and retrofit of modernization; execution of the RAMMP program; and life-cycle support of the F-22A fleet, including supply and maintenance. Finally, the Air Force informed us that it expects to manage future modernization increments as separate acquisitions. However, given the approach the Air Force has taken to date on this and other modernization programs, there is little assurance that this will occur without specific OSD direction.

Recommendation for Executive Action

As new and enhanced capabilities are proposed and vetted beyond Increment 3.2B in the F-22A modernization program, we recommend that the Under Secretary for Acquisition, Technology and Logistics evaluate those capabilities in accordance with DOD policy and statutory criteria to determine if they should be established as separate major defense acquisition programs, each with its own milestones, business case, and cost baseline that includes all applicable direct and indirect support costs required to complete the program.

Agency Comments and Our Evaluation

DOD provided us written comments on a draft of this report. The comments appear in appendix II. DOD also provided technical comments that were incorporated as appropriate. During the agency comment period, DOD requested clarification regarding our recommendation. As a result, we revised the recommendation to more clearly state that the Under Secretary of Acquisition, Technology and Logistics will evaluate future planned F-22A modernization capabilities to determine if those meeting DOD policy and statutory criteria should be established as a separate major acquisition program.

DOD concurred with the revised recommendation.

We are sending copies of this report to interested congressional committees, the Secretary of Defense, the Secretary of the Air Force and the Under Secretary of Defense for Acquisition, Technology and Logistics. In addition, the report will be available at no charge on GAO's website at http://www.gao.gov.

If you or your staff have questions about this report, please contact me at (202) 512-4841 or SullivanM@gao.gov. Contact points for our Offices of Congressional Relations and Public Affairs may be found on the last page of this report.

Michael J. Sullivan, Director
Acquisition and Sourcing Management

Appendix I: Scope and Methodology

To determine the extent to which F-22A modernization met cost and schedule goals and operational requirements, we reviewed documentation of program plans and status, including cost estimates, briefings by program office officials to Department of Defense (DOD) and Air Force oversight officials, annual Selected Acquisition Reports, Defense Acquisition Executive Summary reports, Director, Operational Test and Evaluation (DOT&E) annual test result summaries, Defense Contract Management Agency (DCMA) program assessment reports, acquisition plans, operational requirements documentation, contract documentation, schedules and other data. We reviewed documentation of key decisions made on F-22A modernization, including acquisition decision memoranda and Joint Requirements Oversight Council memoranda. We reviewed F-22A cost performance report data, contract cost data, and budgetary documents. In assessing the achievement of cost goals by the F-22 modernization and other improvement efforts, we compared the program cost estimate from 2004, shortly before development began for Increment 2, with the latest available estimates. We determined what changes in planned capabilities occurred after modernization efforts began. In assessing the F-22A modernization's achievement of schedule goals and delivery of planned capabilities, we identified progress made in delivering new capabilities in accordance with plans, and determined what factors contributed to schedule changes. We interviewed program office officials having knowledge of factors driving cost estimate and schedule changes over time. We also interviewed officials from the F-22A Program Office, DOD test organizations, and Air Combat Command to obtain their views on progress; ongoing concerns and actions taken to address them; and future plans to complete F-22A development procurement and operational testing. We used the latest cost data available during the period of our review; however the F-22A program office was preparing a new cost estimate for F-22A modernization and the estimated costs of increments beyond Increment 3.2B had not yet been determined or added to this estimate.

To determine what progress has been made in completing developmental and operational testing, and resolving system deficiencies, we reviewed DOT&E annual test report summaries and briefings to DOD oversight and requirements officials. We reviewed summaries of recent operational test results provided by Air Force test officials and program risk information related to developmental and operational testing for F-22A modernization. We reviewed documentation of program decisions, including acquisition decision memoranda. We reviewed data from prior GAO reviews on operations and support costs for F-22A and other stealth aircraft, Selected Acquisition Reports, Defense Acquisition Executive Summary reports, contract documents, and program cost estimates. In assessing progress made in operational testing, we compared initial and current operational test plans to determine if significant changes were made after testing began. We identified relevant factors contributing to testing delays. In assessing the resolution of system deficiencies, we identified the number of successful test points flown during operational testing and identified what changes were made in requirements after operational testing began. We determined what key risks and issues remain that could affect developmental and operational testing in the future. We identified issues contributing to increased operations and sustainment costs and to decreased aircraft availability, and actions taken by the F-22A program to mitigate them. We interviewed officials from the F-22A Program Office, DOD test organizations, and Air Combat Command to obtain their views on progress, ongoing concerns and actions taken to address them, and future plans to complete developmental and operational testing. At the time of our review, the final follow-on operational test and evaluation results for Increment 3.1 were not yet available and other test information we had requested was not readily available within the reporting period for this report due to its high classification level. Accordingly, our analysis of actual results and data was somewhat constrained and our reporting limited to providing summary level observations due to the classification level of some of the data. Notwithstanding, DOD officials gave us access to sufficient information to make informed judgments on the matters covered in this report.

In performing our work, we obtained information and interviewed officials from the F-22A Program Office, Wright-Patterson Air Force Base, Ohio; Air Combat Command, Langley Air Force Base Virginia; Office of the Director, Operational Test & Evaluation, Office of the Secretary of Defense, Arlington, Virginia; and the Air Force Operational Test and Evaluation Center, Kirtland Air Force Base, New Mexico. We assessed the reliability of DOD and F-22A contractor data by (1) obtaining and reviewing related information from various sources, and (2) interviewing agency officials knowledgeable about the data. We determined that the data were sufficiently reliable for the purposes of this report. We conducted this performance audit from June 2011 to March 2012 in accordance with generally accepted government auditing standards. Those standards require that we plan and perform the audit to obtain sufficient, appropriate evidence to provide a reasonable basis for our findings and conclusions based on our audit objectives. We believe that the evidence obtained provides a reasonable basis for our findings and conclusions based on our audit objectives.

Appendix II: Comments from the Department of Defense

OFFICE OF THE UNDER SECRETARY OF DEFENSE
3000 DEFENSE PENTAGON
WASHINGTON, DC 20301-3000

ACQUISITION.
TECHNOLOGY
AND LOGISTICS

APR 3 0 2012

Mr. Michael J. Sullivan
Director, Acquisition and Sourcing Management
U.S. Government Accountability Office
441 G Street NW
Washington, DC 20548

Dear Mr. Sullivan:

This is the Department of Defense response to the Government Accountability Office (GAO) Draft Report, GAO-12-447, "TACTICAL AIRCRAFT: F-22A Modernization Program, Faces Cost, Technical and Sustainment Risks," dated March 26, 2012 (GAO Code 120984). The Department appreciates the effort of the GAO and the opportunity to comment on the draft report.

The Department appreciates the collegial, professional working relationship the GAO established with its Department counterparts during this engagement. The Department believes this cooperative relationship resulted in a higher quality and more accurate report, ultimately resulting in a single recommendation to which the Department concurs.

Detailed comments on the report recommendation are enclosed. The Department appreciates the opportunity to respond to your draft report and looks forward to working with you as we continue to improve our acquisition processes.

Sincerely,

David G. Ahern
Deputy Assistant Secretary of Defense
Strategic and Tactical Systems

Enclosure:
As stated

GAO Draft Report Dated MARCH 26, 2012
GAO-12-447 (GAO CODE 120984)

"TACTICAL AIRCRAFT: F-22A MODERNIZATION PROGRAM, FACES COST, TECHNICAL AND SUSTAINMENT RISKS"

DEPARTMENT OF DEFENSE COMMENTS TO THE RECOMMENDATION

The GAO made one "recommendation for executive action" for the Secretary of Defense.

RECOMMENDATION: "As new and enhanced capabilities are proposed and vetted beyond Increment 3.2B in the F-22A modernization program, we recommend that the Under Secretary for Acquisition, Technology and Logistics evaluate those capabilities in accordance with DoD policy and statutory criteria to determine if they should be established as separate major defense acquisition programs, each with its own milestones business case and cost baseline that includes all applicable direct and indirect support costs required to complete the program." [pg 17]

DOD RESPONSE: Concur. The Department agrees with the spirit and intent of this recommendation. In fact, the Department's position is that this is exactly what the Department does and intends to continue doing for each and every potential acquisition program.

Further, the Air Force states they "will continue to structure future modernization programs to achieve optimal cost effectiveness and management oversight, in collaboration with Office of the Under Secretary of Defense for Acquisition, Technology and Logistics, compliant with DoD policy and statutory criteria. However, regulatory and statutory compliance is not the single remedy for cost and schedule growth. Application of accepted program management best practices in technology maturation, systems engineering, test & evaluation, risk and earned value management is also essential to sound program management. The Air Force will continue to employ this combination of policy, statute and program management best practices to ensure that future modernization increments are fielded expeditiously and affordably."

Appendix III: GAO Contact and Staff Acknowledgments

GAO Contact	Michael J. Sullivan, (202) 512-4841 or SullivanM@gao.gov
Staff Acknowledgments	In addition to the contact named above, Bruce Fairbairn, Assistant Director; Marvin Bonner; Sean Seales; Marie Ahearn; Ana Aviles; Laura Greifner; Travis Masters; and Roxanna Sun made key contributions to this report.

Related GAO Products

Tactical Aircraft: Comparison of F-22A and Legacy Fighter Modernization Programs. GAO-12-524. Washington, D.C.: April 26, 2012.

Joint Strike Fighter: Restructuring Places Program on Firmer Footing, but Progress Still Lags. GAO-11-325. Washington, D.C.: April 7, 2011.

Defense Acquisitions: Assessments of Selected Weapon Programs. GAO-11-233SP. Washington, D.C.: March 29, 2011.

Tactical Aircraft: DOD's Ability to Meet Future Requirements Is Uncertain, with Key Analyses Needed to Inform Upcoming Investment Decisions. GAO-10-789. Washington, D.C.: July 29, 2010.

Defense Management: DOD Needs Better Information and Guidance to More Effectively Manage and Reduce Operating and Support Costs of Major Weapon Systems. GAO-10-717. Washington, D.C.: July 20, 2010.

Defense Contracting: DOD Has Enhanced Insight into Undefinitized Contract Action Use, but Management at Local Commands Needs Improvement. GAO-10-299. Washington, D.C.: January 28, 2010.

Defense Acquisitions: Measuring the Value of DOD's Weapon Programs Requires Starting with Realistic Baselines. GAO-09-543T. Washington, D.C.: April 1, 2009.

GAO Cost Estimating and Assessment Guide. GAO-09-3SP. Washington, D.C.: March 2, 2009.

Defense Acquisitions: A Knowledge-Based Funding Approach Could Improve Major Weapon System Program Outcomes. GAO-08-619. Washington, D.C.: July 2, 2008.

Tactical Aircraft: DOD Needs a Joint and Integrated Investment Strategy. GAO-07-415. Washington, D.C.: April 2, 2007.

Tactical Aircraft: DOD Should Present a New F-22A Business Case before Making Further Investments. GAO-06-455R. Washington, D.C.: April 26, 2006.

Defense Acquisitions: Air Force Still Needs Business Case to Support F/A-22 Quantities and Increased Capabilities. GAO-05-304. Washington, D.C.: March 15, 2005.

GAO's Mission	The Government Accountability Office, the audit, evaluation, and investigative arm of Congress, exists to support Congress in meeting its constitutional responsibilities and to help improve the performance and accountability of the federal government for the American people. GAO examines the use of public funds; evaluates federal programs and policies; and provides analyses, recommendations, and other assistance to help Congress make informed oversight, policy, and funding decisions. GAO's commitment to good government is reflected in its core values of accountability, integrity, and reliability.
Obtaining Copies of GAO Reports and Testimony	The fastest and easiest way to obtain copies of GAO documents at no cost is through GAO's website (www.gao.gov). Each weekday afternoon, GAO posts on its website newly released reports, testimony, and correspondence. To have GAO e-mail you a list of newly posted products, go to www.gao.gov and select "E-mail Updates."
Order by Phone	The price of each GAO publication reflects GAO's actual cost of production and distribution and depends on the number of pages in the publication and whether the publication is printed in color or black and white. Pricing and ordering information is posted on GAO's website, http://www.gao.gov/ordering.htm.
	Place orders by calling (202) 512-6000, toll free (866) 801-7077, or TDD (202) 512-2537.
	Orders may be paid for using American Express, Discover Card, MasterCard, Visa, check, or money order. Call for additional information.
Connect with GAO	Connect with GAO on Facebook, Flickr, Twitter, and YouTube. Subscribe to our RSS Feeds or E-mail Updates. Listen to our Podcasts. Visit GAO on the web at www.gao.gov.
To Report Fraud, Waste, and Abuse in Federal Programs	Contact: Website: www.gao.gov/fraudnet/fraudnet.htm E-mail: fraudnet@gao.gov Automated answering system: (800) 424-5454 or (202) 512-7470
Congressional Relations	Katherine Siggerud, Managing Director, siggerudk@gao.gov, (202) 512-4400, U.S. Government Accountability Office, 441 G Street NW, Room 7125, Washington, DC 20548
Public Affairs	Chuck Young, Managing Director, youngc1@gao.gov, (202) 512-4800 U.S. Government Accountability Office, 441 G Street NW, Room 7149 Washington, DC 20548

www.ingramcontent.com/pod-product-compliance
Lightning Source LLC
Chambersburg PA
CBHW081416170526
45166CB00010B/3365